What's Wrong with Math?

A Cure for Societal Math Phobia

Nora M. Tang

ISBN-13: 978-1-989802-36-6

Dandelion Education

Dedications:

For my daughters,

May you one day find yourselves in an elephant-less room surrounded by a boundless field of dandelions.

Table of Contents:

Acknowledgements:

This book was made possible by the help and feedback from the following friends and family, in alphabetical order by last names:

Victor Cheng

Susanna Du

Dr. Susan Gerofsky

Dr. Rebecca Ko

Joseph Liu

John Smith

1.

A Mathematician's Lament and the Elephant in the Room

I'm a math teacher at heart. I've been a math teacher at heart since I made the decision to become one in Grade 10. No math teacher is a stranger to the concept of math phobia. I don't mean the medical condition known as Math Anxiety although that is certainly related to our topic here. I mean the attitude of the general population that math is somehow harder, scarier, less fun, and less practical than any other subject taught in school.

I believe this is quite a universal experience for all math teachers. When we are asked what subject we teach and we answer "math", we would often get a response like this:

"Oh wow! You must be smart!"

I understand this to be a social compliment, aimed to show the listener that the speaker is friendly and open to forming a social bond, except it's often followed by something like this:

"I can't do math for the life of me. I can't even add five and seven without a calculator."

I understand that this is also a social gesture. Exposing our weaknesses shows vulnerability, and vulnerability fosters trust, an important building block of all human relationships. What bothers me is the fact that it is somehow only socially acceptable to expose this level of weakness when it's related to math and no other subject. If we admitted to not knowing how to write the word "red" in our dominant language or how to tell the difference between an animal and a plant, we would receive at least a look or two of concern for our inability to function as adults in society. It's only when such an admission is made about math that people respond with empathetic pats on the back and softly muttered agreements that math is indeed the devil of all subjects.

It's all very strange if we sit down and think about it. The amount of information a student has to process on a regular basis in language classes is astronomical compared to in math classes. Take reading a paragraph for example. A student would have to parse the text into letters, merge the letters to form phonic sounds, blend the phonic sounds to form words, recall the meaning of each word, piece the meanings together with their corresponding parts of speech to understand the sentence, and relate the sentences to each other to understand the paragraph as a whole, all the while paying attention to punctuation, verb tenses and subject-verb agreement, point of view, voice, and sentence structure. This is only the most basic understanding of the text. Then they have to infer the unspoken messages between the lines, identify and understand literary devices like metaphors and symbolism, extract summaries and themes and morals, and apply them to questions beyond the text itself. It is a huge amount of mental work, yet students do it every day without batting an eye.

We can compare that to what we typically require of a student in math. A student would read two to three lines of a "word problem", pick out two or three numbers and figure out where they go in a *given* formula. (We don't even mix up the formulae in fear of confusing students!) Then the student would punch them into a calculator and record the answer and call it a day. Yet, people would overwhelmingly choose to read a paragraph over doing a math question all the while admitting to their fear of math.

Despite my fortunate exposure to the much more interesting world of math competitions while I was growing up, I was still under the impression that the unattractiveness of "school math" lies in characteristics of mathematics itself, that math itself makes for a boring, and to many people, scary, subject to teach and learn. I deeply understood the apathy because "school math" was boring even to "math kids" like me. It was similar to understanding if someone didn't want to plant dandelions in their garden. They are easy to grow, but the reward they would reap is unimpressive. The fear, on the other hand, always had me scratching my head. What's there to be afraid of in a dandelion? It doesn't even have thorns.

What's there to be afraid of in math?

It's not until recent years, when I had the opportunity to watch my own children learn in their early years, that I began to question our tendency to overly differentiate among the subjects and people's ability to learn them. I became more and more convinced that it's not the basic mathematical concepts and skills themselves that are boring and dry and scary and hard for a seemingly large proportion of people who were not "born with the mathematical aptitude". It's not the math.

It's not people's natural inability to learn math either. It's the way we as educators teach math that's the problem, and that problem comes from the way we as a society think of math.

I came across Paul Lockhart's *A Mathematician's Lament* (circulating in a PDF file at the time, before it was published years later) during my university years. For those who are unfamiliar with the title, it's essentially a thirteen-thousand-word complaint about how Mathematics as an art form is misunderstood by the general public and, as a result, mistreated by the school system. (P. Lockhart, 2009) It hit me like a ton of bricks. By then I had long since found my calling to become a math teacher, full of passion and aspirations for both math and teaching. As I read my way through it, I cried tears of bitterness at the injustices the art of mathematics suffers at the hands of societal misinterpretations. How dare they reduce such a magnificent and elegant art form into an unattractive, complicated, boring, scary, and seemingly useless *tool*! I thought anyone who cared enough about the subject and the sacred profession of teaching to become a math teacher would share my resonance with Lockhart's message, but when this PDF file made its rounds into our local math teachers' Listserv, the responses shocked me once more.

There were a few who expressed their agreement with Lockhart in the discussions that followed, but the overwhelming consensus was that students didn't enjoy the math they learned in school because we weren't "connecting it to real life." Frankly I was under the impression that I, a young adult taking part in a discussion about the education of our future generation, hadn't even started my "real life" yet. (I was still in school and living with my parents.) Not only were the "real life" examples brought up in the discussion things that I hadn't encountered, things like mortgages, credit card debts, and investment growth, they sounded more like headaches to me just thinking about the possibility of having to deal with them in the future. I highly doubted that a high school student would find them interesting.

I'm not saying all real life references are uninteresting, but just take a moment to imagine the comparison: In a physics class the child learns about electro-magnetic waves and the visible light spectrum, and the teacher talks about its applications in searching for the edge of the universe. In social studies the child learns about the democratic voting process and when and how they can begin taking part in this duty that shapes the course of our country. In biology she learns about photosynthesis and its role in providing food source for the entire ecosystem. In English she learns about the power and elements of persuasion that will help her in every interpersonal situation in life. Then she comes to math class, and the best we can come up with is how to calculate mortgage and credit card debt, which the bank and credit card company would do for her on her bill anyway. Now I'm beginning to see why people hate math.

But wait, you say, *that's <u>not</u> the best we can come up with in math. We teach logical thinking. We teach problem solving. Those are plenty useful in life.*

Yes, it would be nice if we actually teach those because math does have those to offer. Do we though? Do we really? Where are the logical thinking and problem solving when we drill students on the process of long division? Where are the logical thinking and problem solving when we assign 30 questions for homework on the area of circles that have students plug in the exact same formula in the exact same way with 30 different sets of numbers? Where are the logical thinking and problem solving when we have students memorize each step in solving multi-stepped equations?

The problem is that most of us don't know what logical thinking and problem solving look like anymore. We can see this in our provincial math curriculum. Logical reasoning is a tiny chapter in Foundations of Mathematics 11 (https://curriculum.gov.bc.ca/curriculum/mathematics/11/foundations-of-mathematics), one of three pathways by this grade level, and the other two pathways have no mention of it. The phrase "problem solving" is either a hand-waving, overarching brush that paints over the whole curriculum, so broad that no one knows what to do with it, or it's such an afterthought that it gets attached to a specific skill, like "applying […] in real-life contexts and problem-based situations." (https://curriculum.gov.bc.ca/curriculum/mathematics/5/core)

With that kind of instruction to teachers and textbook authors, no wonder we end up with the kind of "problem solving" at the end of each chapter exercises.

The student does 20 questions that look like this:

Find the area of a circle of radius 5.

Then she gets to a "problem solving" question that looks like this:

A circular dinner plate has a radius of 5cm. If Peggy the plate painter charges $1 for every square centimeter of painting, how much should she charge to paint the top side of this plate?

Where exactly is the problem solving in this? Is it the fact that it's about a plate and not an idealistic circle? Is it the fact that it contains a person's name? Is it because it is a "real life situation"? You and I both know we won't run into this problem in life. In fact, it doesn't matter if this problem actually exists in real life. The problem with this "problem" is that it's not a "problem" at all. It's a drill, just like the 20 before it, dressed in a problem's outfit.

You see, our current system of math education is very efficient at creating math phobia. This systemic math phobia is the elephant in the room. Everyone knows it's there. Everyone knows it shouldn't be there. No one knows how it got there or how to get it to leave. In the end it gets left in the room year after year, decade after decade, generation after generation. Occasionally someone would come along and give it a tentative nudge this way or that without much of a reaction. Then once in a long while someone like Lockhart would come along to try to convince others on what we are missing, but those voices would make their ripples before disappearing again. It's because as a whole we haven't seen nor experienced what we are missing. As a whole we don't know what the room looks like without the elephant.

Therefore I am writing this book from my corner of the room, to simply tell people about the part of the elephant I see from here, to lend my feeble voice in painting a clearer picture of what it is that we're missing, what it is that we need to do to move the elephant, and a suggestion or two on the ways we might get it done. My only goal is to give whatever I can offer for the future of an elephant-less room for all, and I invite all who are willing to help. Moving this elephant needs an orchestrated effort from a great many people.

2.

Clues from My "Idle" Years and the Four Levels of Learning

What exactly are we doing wrong in math? What is it that's missing from math education that's not missing from anywhere else? Are we really missing that sense-making and meaning-seeking element that is part of this new Core Competencies curriculum? Is that our silver bullet? There's no doubt that we are missing meaning of some kind from our math curriculum, but this "going back to the basics" type of meaning-seeking just did not feel right to me. The question of what kind of meaning we need in math stumped me as much as it stumped many others, and I left it in the back of my mind for many years.

Then I quit my job as the Math Department Head at Wuhan Maple Leaf International School, which offers the BC curriculum to senior high school students in China. I put a pause on my career to start a family back here in Vancouver, and it ended up being a decade-long voluntary house arrest. I didn't thrive in motherhood nearly as much as in my job as a math teacher, but it did give me the time and incentives to ponder about a few things.

Clues from Chinese Flashcards

When I was pregnant with my first child, I worried constantly about the prospect of her growing up unable to read, write, or maybe even speak Chinese. She'd have to whip out Google Translate just to read a text from one of her grandparents. She'd be like that family friend my mom told me about, where the mother asked her son to translate a letter from the government, and the son didn't know enough Chinese to relay the message after reading it. I searched franticly online for information on how to raise a bilingual and biliterate child in a monolingual society.

After realizing that there's no systematic teaching material available here in North America that can support a child's Chinese learning until literacy, I decided to sit down and make my own flashcards and use the local library for reading resources. How many did I have to make? As it turns out, I didn't have to make very many at all.

The definition of literacy is a reading level of around Grade 8, by which students who live and attend school in China have an average reading vocabulary of 5500 Chinese characters. However, a child can begin reading lower level books independently with as few as 1000 commonly used characters, or the vocabulary level of a Grade 2 student. Now, here is the eye-opening part: It's recommended that the remaining 4500 characters (or anything after the initial 1000) be picked up through self-directed reading instead of explicit instruction.

By the way, this is not just limited to Chinese. It's quite common in languages around the world where students can begin independent reading with a vocabulary level of around Grade 2, after which the easiest, quickest, and most efficient way of improving language proficiency in terms of vocabulary or grammar is not through direct instruction on vocabulary or grammar, but through reading.

Hmm... That's interesting, I thought to myself.

Clues from Piano Drills

We bought a piano when my younger child was born. It was a longtime dream of mine to own one and to learn to play properly, so when my little monsters were old enough to leave me alone for half an hour at a time, I practiced. I began with Hanon's *The Virtuoso Pianist.* For those who are not into learning piano, it's a standard book of drills for beginners designed to build strength and agility. It was simple, repetitive, mind numbing to play, blunt noise to listen to, but it served its purpose. After about a month of drilling, I could easily sight read songs that I had difficulty playing before, now that my hands could keep up more easily with my eyes.

When my older one turned 4, my mom suggested that I look into introductory piano lessons for children. Naturally as a teacher who played the piano, I wasn't willing to just sign her up for some lessons and call it a day. Again, I searched endlessly on the web for information. What's the best age to start? Should she learn sight reading or ear training first? Which textbook series is the most pedagogically sound? When is it best to start her on Hanon?

The answer to that last question, to my surprise, was never. No, don't start her on Hanon, *ever*. Many blogs and piano instructor forums I came across suggested that the drills like Hanon do more harm than good in the grand scheme of things. Instead, they suggest that children develop their techniques by learning increasingly difficult pieces of music written for the purpose of performance, as opposed to exercise.

Hmmm... That's interesting, I thought to myself again.

Clues from Art Lessons

Then one day my older one expressed interest in learning to draw. Once again, I searched the internet for a tried-and-true, widely accepted training routine for beginning artists. Well, aside from the most basic things like how to hold a pencil and how to draw straight lines, circles, dots, and other types of marks, *there is none*. There are targeted practices, like how to draw hands, how to draw flowers, proportions, shading, colours, perspectives, but those are highly goal-specific, largely dependent on what the artist wants to be able to do next. There's *no* training "course" that everyone does to bring them from beginner to mastery. The consensus is that you're supposed to jump in and draw something you see and like each day. It doesn't matter if it's from other people's drawings or photos or real life. What's important is that you keep drawing. You'll know what specific things to practice and develop your own style as you go.

Hmmm... That's interesting, I thought once more.

I think I'm beginning to see what we're missing in math here, but before I get to that, let's go over what I think are the 4 levels of learning in a field of study. I see these 4 levels in many other fields, but when it comes to math, we seem to be missing a huge portion.

Semantic Learning

This is the first impression phase when coming into contact with a new field or a new concept in a field. It contains information about how the very basic mechanics work. For example, in English, this would include realizing that the direction of text goes from left to right, and that the lines go from top to bottom. This also includes figuring out that reading doesn't involve saying the name of each letter you see, but parsing words into phonic sounds, then blending the sounds to form syllables. In visual art, this includes learning to use drawing tools and getting used to the idea of drawing what the eyes see. This level of learning is usually quick and simple. Once the child crosses the threshold of understanding and becomes oriented in the subject, there's not much else to do here.

Technical Learning

This is where the bulk of the "hard work" happens when we think about learning. In the example of English, this would include learning to write all 26 letters in upper and lower cases, committing the phonic sounds and their letter combinations to memory, learning vocabulary through memorization, rote reading/singing, flashcards, repetitions, and memorizing syntax and grammatical rules. In piano this would be the finger drills, scales, arpeggios, chord exercises, and memorizing which dot on the grand staff corresponds to which key on the keyboard. The main characteristics of this level of learning in any subject are that it's mindless, boring, repetitive, and when not given a clear goal, meaningless and pointless.

Does that sound familiar to you?

Contextual Learning

In most other fields of study, this is where learners spend the majority of their time. It's where the learners engage with the creations of others in the field, studying them, absorbing them, reflecting on them, and appreciating them. In English, this is reading works that others wrote. In piano, this is learning to play pieces of music that others composed. In visual art, this is imitating artworks that others created. Even in sciences, which we might think are dominated by memorization of facts, learners spend time studying the theories, experiments, and results that others conceived, designed, and obtained. They study what conjectures past scientists thought up, what they did to test their conjectures, what conclusions they came to, and why. Frankly, they do quite a bit more of that logical thinking that we claim to do in math. This level of learning is how "meaning" is passed from one person to the next. It's how we communicate with strangers on a deeper level of significance.

How often do the average K to 12 students do this in math? They don't. That's the problem.

Creative Learning

This is when the learners absorb enough of other people's work and capture their own inspirations to give back to humanity's pool of intellectual wealth in the field. They elaborate on their inspirations in a form that's widely accepted in the field to share with others. It doesn't have to be groundbreaking or revolutionizing at first. It doesn't even have to be very interesting at first. It's a learning process that tends to alternate between *Contextual* and *Creative learning*, and it can be a lifelong process to mastery. The examples are pretty straight forward. In English this would be writing. In music, it's composing or songwriting. In art, it's creating artworks. This is the level of learning that gives human beings the sense of purpose because we're contributing to society in a meaningful and lasting way. We're making the world a slightly better place than when we found it.

We have an innate drive to create and to contribute to the greater good, and right now, our math curriculum disregards that drive.

But wait, you say, *isn't this a simplified/modified/reclassified version of Bloom's Taxonomy?*

I hear you, my fellow teacher. The decision to call those 4 categories "levels" like we do in Bloom's Taxonomy was my deliberate choice. The short answer is no. Let's set straight a couple of things about Bloom's Taxonomy first.

First of all, let's clarify that Bloom's Taxonomy, as the name suggests, is intended to be a system of classification, not prescription. It consists of 3 domains: cognitive, affective, and psychomotor, with the cognitive domain being the first to be introduced and the most widely used by educators. The cognitive domain contains 6 levels of learning goals:

Knowledge, Comprehension, Application, Analysis, Synthesis, and Evaluation.
(Bloom et al., 1956)

Bloom's Taxonomy's main purpose is to classify assessment questions so that teachers can communicate more effectively. The levels mentioned here refer to the types of questions a learner is able to answer using what he knows about a particular concept. For example, regarding the concept of compound sentences, a learner who has trouble moving beyond questions of the first level (Knowledge) might only be able to recite the definition of a compound sentence without complete understanding of what it means. A learner who can answer the third level (Application) questions might be able to put two simple sentences together to form a compound sentence. Is it possible that a learner answering third level questions correctly is unable to recall the precise definition of a compound sentence in words? Yes, but he would still have a better understanding of the concept of compound sentences than the first learner.

What Bloom's Taxonomy does *not* tell you is that you should teach every concept starting with the memorization of its definition and progressing through each "level". It doesn't work for every concept, and it's not meant to work that way. It's not a prescription for the order of learning, and therefore teaching. Yes, as teachers we aim to bring a learner currently at a lower level to a higher level for the concept we are teaching, but we don't have to do that by intentionally hitting every level along the way in the given order.

Secondly, math teachers, textbook authors, and curriculum designers seem to think of Bloom's Taxonomy in a "singular" way, meaning as we move to higher levels, we are still talking about the same single concept. This is exactly how we end up with "problem solving" questions like the plate painter one from last chapter. The problem is that the more fundamental and technical the concept is, the harder it is to design questions or activities that demonstrate its use in isolation at a high level. How do you get a learner to demonstrate the skill level of "Evaluation" with only the concept of compound sentence? How would that differ from demonstrating "Evaluation" with complex sentences instead? You don't, and it wouldn't. The higher the level, the more it requires the student to demonstrate concepts in batches, like "using a variety of *sentence structures*, correct *punctuations*, *paragraph structures*, consistent *voice*, *verb tenses*, and demonstrate knowledge of the *audience* and *elements of persuasion*" instead of "using *compound sentences*".

English teachers know this. They don't try to get learners to demonstrate using every word in a sentence before letting them read it in a story. They don't stubbornly follow every concept up the ladder of Bloom's Taxonomy to the bitter end. They have no trouble letting kids run before they can crawl steadily because they know running, stumbling, and getting up to continue running is the most efficient way for kids to become proficient crawlers, walkers, *and* runners in English.

My daughter's kindergarten class was practicing writing full sentences as soon as they finished learning phonic sounds. At the time they had learned to spell exactly 0 words. The sentences they produced were nowhere near complete. It was rare to find more than one word correctly spelled, or even phonically accurate in a sentence they wrote, but they were writing sentences of their own creation none the less. They were doing *creative learning*, with a side focus on trying to write down sounds they hear as letter combinations.

The reason we don't see this in math is not exactly the fault of teachers, or textbook authors, or curriculum designers. We never let our students do this in math because no one knows what running looks like in math. Our students are stuck in *technical learning* year after year because we don't know when to stop and what comes after. We don't know what the *technical learning* is *for*.

With that in mind, let's take a closer look at the relationship between *technical learning* and what supposedly comes after in the field of mathematics.

3.

The Magic of Contextual Learning and Our Dire Situation in Math

About a year before my older child started school, my mom decided to teach her some math during her weekly visits to see the kids. She would play rock-paper-scissors with my daughter, and they would keep track of the score mentally by adding or subtracting 1 each time. Then she moved on to adding by 2s by playing two-handed rock-paper-scissors. This way when one person won a round, she won two points instead of one. My daughter would take her time counting up or down with her fingers.

One day after one of these teaching sessions, my daughter came to me and asked, “Mommy, what’s 5 plus 3?”

I was washing dishes at the time and told her to figure it out with her fingers. To my surprise, she did. To make sure she was really doing addition and not guessing, I gave her a few more that are within ten and a couple slightly above. She did all of them with the help of her fingers and occasionally toes.

I was excited of course. After all, we never formally taught her addition yet. I called my mom to tell her the good news, but after listening to my recount of the event, my mom promptly concluded that it wasn’t addition.

"She has to use her fingers," my mom replied on the phone. "That's cheating. She's not doing real addition."

If a child could play a recognizable melody on the keyboard without the help of music scores, we would say she's talented. If a child could spell out words she never memorized by using phonic sounds, we would celebrate. It's only in math where we have trouble recognizing abilities that are not the direct result of prolonged drills, memorization, repetition, abilities that are not the direct result of *technical learning*.

Technical learning has its place. Hanon's drills helped me play some pieces of music I couldn't play before. However, *technical learning* is always supposed to come with a goal, an end point. All the drilling I did in Hanon had to stop to make way for learning more interesting pieces of music, for *contextual learning*. In teaching Chinese we use flashcards to build up vocabulary until the child can read independently. When we reach that point, the flashcard drills end to make way for reading sessions, for *contextual learning*. When we train basketball players, all the dribbling, passing, shooting, lay-up drills have to stop at some point to make way for an actual game, for *contextual learning*. Without the game, there is no point to all the drills. The goal of *technical learning* is to get us to the point where we can do *contextual learning* without too much difficulty.

But wait, you say, *don't professional basketball players still do drills? One of the most inspirational messages a professional in a field can give to a beginner is that he is still doing basic drills every day.*

Yes, of course, but those drills still come with a goal. The goal is always to improve some aspect of the actual game, of *contextual learning*. Doing drills aimlessly just doesn't happen in other fields.

The problem in math is that we don't know what we're drilling for. When our students complain about math being useless, they're not talking about using it in the real world. What they really mean is that it's meaningless. We don't ever use it in a meaningful and significant way that resonates with our students. We don't know what *contextual learning* looks like in math.

If in other fields *contextual learning* is about studying the works others have produced, isn't that what we're doing in math when we teach the number system? Isn't that what we're doing when we teach methods for solving systems of equations? Formulae in geometry? Notations in calculus?

Here's the big misconception of what mathematicians do. The creative works that mathematicians produce are not in the solutions and notations and formulae. The creative works that mathematicians do are in the problems they think about. They make problems, and they try to solve the problems they make. The notations and systems and formulae they create out of that process are just byproducts, tools to help them understand their problems in new and clearer ways. Focusing only on the formulae and notation is like throwing out a novel and keeping the plot outline, like throwing out the painting and keeping the colour palette, like throwing out a piece of music and keeping the scale it's built on.

One of the centerpieces of Lockhart's (2009) message in *A Mathematician's Lament* was that math is really about thought experiments. In fact I would go as far as to say that all creative works are thought experiments. The magic of *contextual learning*, what makes it meaningful and worthwhile, is that we are engaging with another human being's thought experiment, played out and captured in physical form. The significance comes from following another human being's thoughts on something they find meaningful enough to capture. The resonance comes from finding traces of the human being, who is similar to us in some way, on the other side of the work. The interest comes from appreciating the ingenuity of the human being whose idea it was to conduct this thought experiment.

Of course we won't find every problem we come across interesting. We won't resonate with every story we read or like every song we learn either. It doesn't change the fact that we will keep reading and learning new music because we know that the moment of connection we feel when we find a piece of work that speaks to us is worth the search in between.

But wait, you say, *the "artsy" subjects in these examples, like languages, music, and visual art, allow for more freedom in exploration and subjective expression because they're "artsy". Math, being the foundation of quite a few other "scientific" subjects, has to focus more on accuracy and objectivity. Wouldn't shifting this focus to the more free explorations associated with* contextual *and* creative learning *cause damage to those fields of science down the road? Wouldn't that make math incompatible with* contextual *and* creative learning*?*

I'm glad you brought up this dichotomy between the Arts and the Sciences because I'm now a firm believer that the division is not only made up but actively impeding the advancement of some fields of study. Until we clear up this misunderstanding about the subject of mathematics, this book is unlikely to have any real impact on the way people do things in math education.

Let's talk about the notion that Mathematics is about accuracy. I guess the notion comes from the fact that most sciences are based at least partially on calculations, and that calculations are a major focus for the majority of the K to 12 math curriculum. However, that is not to say that calculations and accuracy are therefore what math is good for or what math is all about, or even that we should therefore focus on calculations and accuracy in math education. Let's try making similar statements about English, for example, to see what we're talking about.

We know that legal documents are written in English, and inaccuracy in grammar used in legal documents can have serious consequences in the legal system. Does that mean English taught in school should take on grammatical accuracy as its main focus? Does that mean English as a subject is about the precise correlation between the words being used and the meaning intended? Does that mean it's wrong to say that English is a tool of creativity that can be used to express feelings and communicate big ideas? Does it also mean that if we focus on creativity instead of grammatical accuracy in English we would cause the collapse of our legal system?

No, we wouldn't cause the collapse of our legal system. This is because precision in legal language can be taught separately in law school, just as numerical accuracy can be taught separately in the branches of science that require it.

The reason why we have no problem accepting "artsy" descriptions of English, like "creative" or "open ended" or "explorative" or "expressive", is because we as a society have seen English used in contexts other than a legal document. We know that English can be fun. The reason we have such a problem accepting the same descriptions of math is because we as a society haven't seen that side of math. Few people ever have the chance to work through a math puzzle for fun during their entire school experience, and far fewer ever have the chance to make a math problem to wrestle with for the kick of it.

That's nice, you say, *but we spend 120% of our class time doing basic drills, and my students are still struggling. We don't have time for explorative problem solving for fun.*

Yes, I believe you, and I'm not surprised. However, I would suggest that it is precisely *because* we spend all our time doing basic drills that our students perform at such a low level.

According to an article published by *The Economist* in 2013, the average adult native English speaker has a vocabulary range of about 20,000-35,000 words. (R.L.G., 2013) Can you imagine what our language proficiency level would look like if English teachers taught all those words by giving out vocabulary lists with definitions to memorize instead of through reading? If they taught one new word a day, seven days a week with no holidays all year round, 20,000 words would take close to 55 years to teach, not to mention the amount of vocabulary review they would have to do each day.

Remember constructivism (Piaget, 1936), which says that learners constantly modify their own definitions and ideas to fit new experiences. We humans are not the logical beings we think we are. We learn much more effectively when being submerged in context than when building from the definitions up in a vacuum.

Contextual learning is not just an "interest booster" to motivate students to learn the subject. *Contextual learning* is much more efficient in the long run at teaching large sets of skills. This is because *contextual learning* provides huge numbers of opportunities for students to practice basic skills without making it feel like practice at all. Take Chinese vocabulary building for example. Reviewing 25 flashcards of Chinese characters a day takes about 15 to 20 minutes, and it can feel tedious when the characters are forgotten and have to be relearned. However, in the course of a 15 to 20 minutes reading session, the student easily reviews over 100 characters without it ever feeling like reviewing or relearning forgotten characters. Not only is it time efficient, the context of the story makes it easier for the student to remember the characters themselves, and it's also much more fun to read a story than to look at isolated characters without context.

The reason we need *technical learning* at all is to get over the bumps of difficulty in *contextual learning*. Let's not do it backwards.

Now that we have a vague idea of what we're looking for, let's move on to the next problem: We don't have authentic problems. As a society, we are so alienated from the idea of creativity in math that we don't do it as a hobby. As a result, our reservoir of mathematical thought experiments made by curious human beings is pretty much empty.

You see, it shouldn't be the math teacher's job to come up with problems as they teach. It sure isn't the English teacher's job to write every book they want their students to read, nor is it the music teacher's job to compose all the music they will study in class. In each field there is an ever growing pool of creative work readily available for teachers to use in their classrooms, and ours is empty.

If *semantic learning* is planting a seed, and *technical learning* is the fertilizer that temporarily boosts growth, then *contextual learning* is the rich soil that facilitates long term growth. Right now, we are planting our dandelions on bare rocks in math. We don't have good soil. No wonder our dandelions die and turn into elephants.

Oh, but we do have problems, you say. *There are hundreds, if not thousands, of math contest questions being made each year by contest organizers for K to 12 students around the world*

You are right, and these contest questions are a pure blessing to have and use when we have nothing else. They are infinitely better than the plate painter "problems" we have in textbooks, but they are still not ideal for our purpose for a few reasons.

First of all, they are designed to separate students who are already good at math drills. The lower the grade level, the more this seems to be true. In Grade 4, when these contests typically start, almost all the problems are aimed to test computational fluency instead of being asked out of genuine curiosity.

Second, they come with a time limit and an answer key, which often contains only the final answers. This makes it very hard for teachers and students to not think of the speed and accuracy of the final answer as the most important parts of the solution. In fact, they should be the least important parts, almost afterthoughts, if what we are trying to solve are real problems and not drills in disguise. The most important part should be the *approach to* and the *wrestling with* the solution instead.

Third, they are designed with the solution in mind. This makes them formulaic, especially the ones designed for lower grade levels. Students who do enough of these will start to get a sense that they are not solving new problems, but the same few old ones with new numbers. It's the feeling of "I've seen this trick before," similar to the feeling of reading a story with an unoriginal plot line.

Fourth, they don't inspire those who solve them to come up with their own problems. This is evident in the fact that we now have generations of students who have gone through these contests, but we see no increase in recreational problem making. This might be because of the lack of awareness of problem making as an option for a hobby, but it might also be the way contest questions tend to be presented, with only the name of the organization hosting the contest and not the name of the person who came up with the idea for the problem. This gives the impression that only authority figures in the field of mathematics can make problems, which is neither true nor helpful for our problem shortage situation.

In short, contest questions are a great start into the world of problem solving, and many of them offer much potential with a bit of modification. However, giving them as they are to students or using them alone might not fully achieve our goals of mathematical contextual learning. Thus our problem remains that we have an extreme problem shortage in the field of mathematics.

Now that we are clear on the root of our elephant problem (no pun intended), let's roll up our sleeves and start working on repairing the system.

4.

What Good Problems Look Like and How to Make Them

Let's make one thing clear before we start. I have no idea what a good (popular) problem looks like. No one does in this day and age because this is an uncharted territory. Most people who have had some experience working with math problems will have an idea of what makes a good problem for themselves, and there do tend to be some common themes that many people appreciate in their problems. We will talk about those throughout this chapter. For the general population though, no one has had the chance to do any market research yet. Traditionally, contest hosting organizations make problems, and students solve them. If the students didn't like the problems, that's just too bad for them. Therefore the characteristics of a problem that *sells* to people at every level of mathematical proficiency still need to be tested out.

Let's also be clear on the fact that we will probably make a large number of bad (unpopular) problems in the beginning. It's the same as if we were learning to write stories or compose music. Our first works will most likely be shots in the dark. Since as a society we have no experience in this field, it will probably take a while before we come up with any guidelines for good problem making or classifications, genres, difficulty indication systems. It doesn't mean we're doing it wrong. It just means there is still progress to be made, and we are making it and learning along the way.

The intent of this chapter is not to provide you with tried-and-true methods of generating good problems on demand. The intent is to give you a few sources of inspiration to get you into the flow of making your own problems. Once we have a large enough pool of problems available for use, we can then sort them by difficulty, interest, or curricular needs. For now though, anything you can make helps in both adding to the pool of problems and promoting the idea of problem making as a hobby, so don't worry too much about sticking to a particular topic or difficulty level.

With those in mind, let's start with a couple of ways I have used to find inspirations for problem making. Then we will talk about some methods other people have used.

The first method is to ask and wait for inspiration to come. I know what you're thinking. *When has that ever happened?* Right? The thing I find with inspirations is that it comes only when I am actively looking for it, but there is always a delay. If I want to write a story, I would have to sit there and try to come up with an idea, and I would draw a blank in the moment. I still have to do this though because if I do, I would get an idea or two in the next few days when I am focusing on other things in life. This is because our minds are more open to our surroundings to pick up inspiration after we actively seek it.

I first experienced this with a math problem idea when I was on a trip during a summer break of my university years. We had a family friends' gathering a few days before, and we played this silly but long game of fortune telling with a deck of regular playing cards. Part of the idea of the game is that we shuffle the cards and keep taking the top and bottom cards in the deck to see if their numbers matched. If they did, we added the pair to the list of cards we use to do the fortune telling, with each number representing a different question and each suit representing an answer of some sort. If the top and bottom cards didn't match, which was too often the case, we added them to the discard pile.

I wasn't all that interested in the fortune telling part, but the discarding process kept coming back to me while I rested in my hotel room during the trip. It was difficult to collect all pairs of cards we needed, and we had to keep reusing the discard pile when we ran out. Was it possible for us to be in a loop of discarding cards and reusing the discard pile, never to find another pair of matching numbers? How many times would we go through the discard pile to complete the loop?

There is our inspiration. Now we just need to define a few things to make it a problem.

> Suppose we have a deck of cards numbered from 0 to n, and initially they are in numerical order in a pile from the bottom to the top like this:
>
> 0, 1, 2, 3, … n
>
> For simplicity's sake, let's say the numbers are printed on both sides, and that there is no difference between the two sides. Also, when we "shuffle" using the process above, we don't turn over any group of cards. This prevents us from changing the order of the cards in other ways. If we

keep taking the top and bottom cards and adding them to the discard pile, when we run out of cards, the discard pile would look like this:

0, n, 1, n-1, 2, n-2, …

(Notice the card with the number 0 is always on the bottom, hence my choice to call it 0.)

The question is for any given natural number n, can we find out how many iterations of this shuffling process we have to go through to return to the order of the original deck?

For clarity, let's try this for n=6

Iteration 0 (starting):	0, 1, 2, 3, 4, 5, 6
Iteration 1:	0, 6, 1, 5, 2, 4, 3
Iteration 2:	0, 3, 6, 4, 1, 2, 5
Iteration 3:	0, 5, 3, 2, 6, 1, 4
Iteration 4:	0, 4, 5, 1, 3, 6, 2
Iteration 5:	0, 2, 4, 6, 5, 3, 1
Iteration 6:	0, 1, 2, 3, 4, 5, 6

Now we are back to where we started, and it took 6 iterations.

This is the question about the length of the loop. We can also ask questions about the possibility of avoiding certain combinations of top and bottom cards. If we feel like it, we can also modify the shuffling process and ask the same questions. We would get a different problem.

It has been years, and I would take this problem out to think about it whenever I'm in the mood. I haven't solved it yet, which is fine and doesn't make it any less interesting of a problem. One of the minimum requirements for a good problem is that I haven't solved it yet. If I have solved the problem before, it wouldn't be a problem to me anymore.

In fact, it is often noted by experienced problem solvers that if a problem doesn't require the solver to try and fail at some approaches, if the solver can "see through" the problem in terms of the steps required to arrive at a solution as soon as she reads it, the problem is too easy and therefore not a good problem for the solver. We as a society have to get used to the idea of the process of problem solving being more important than arriving at the right final answer. Even an incomplete solution, when it's the result of active wrestling with the problem, is more valuable of a learning experience than a complete solution to a problem that doesn't require any wrestling.

In the book *Thinking Mathematically*, Mason et al. (2010, p. 45) talk about the act of wrestling with problems (which they call the state of being STUCK) in great detail. They talk about how to think and feel about being in such a state as well as things to try while in this state to move forward. If you're more interested in developing the right habits and attitudes in the problem *solving* side of mathematics, I highly recommend this book.

Now let's talk about my second, and perhaps faster, way of making problems. It is something I learned from the creative writing community when I had a hobby of writing stories. The idea is to take two separate concepts, let's say a square and an equilateral triangle, mix them together, and ask a question about it, like this:

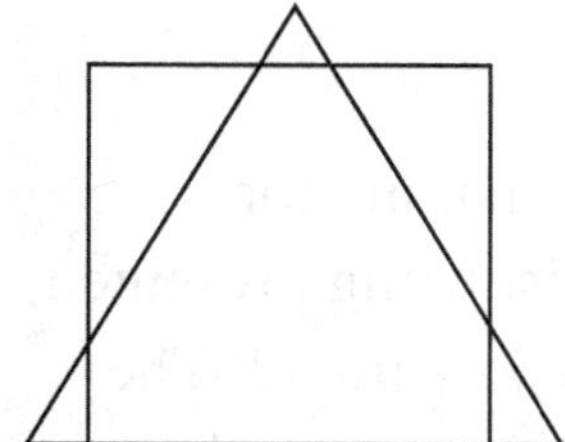

Given the side length of the square (s centimeters) and the side length of the equilateral triangle (t centimeters), calculate the overlapping area. (Figure not drawn to scale)

Of course we still need to define a few things about it, like the fact that the square and the equilateral triangle have their base sides overlapping, and that they share the midpoint of their bases. This is still much faster than waiting for an inspiration to strike, and if we don't like the first mixture of ideas, we can just find another pair and repeat the process. Again, I haven't given any thought to what the solution might look like while making this problem, nor have I considered what grade level this is for.

This brings us to the second trait of good problems that experienced problem solvers tend to agree on. A good problem isn't usually presented as directly linked to a specific skill (part of the curriculum or otherwise). Part of the fun of problem solving lies in figuring out and deciding for ourselves which tool(s) in our toolbox of skills will work on the problem. It's akin to the difference between looking at a structure and asking ourselves "How can I build this with the tools in my box?" as opposed to following a step-by-step building exercise that practices the use of a hammer.

In *Thinking Mathematically*, Mason et al. (2010, pp. 120-132) also dedicate a chapter to making (or *noticing*, as they call it) good mathematical problems. They specifically talk about extending existing problems to make them more mathematically interesting. Here is an example:

Let's go back to our plate painter "problem". Now imagine for a second that you are a professional plate painting artist being presented with a blank circular plate. What would you like to do with it? The important thing to keep in mind here is to throw out the notion that this is an "Area of Circles" problem designed for the Circles chapter in the Grade 7 curriculum. We can explore and find pleasant surprises only when we refuse to begin with the solution in mind.

Perhaps you really do have a fondness for circles and decide to paint a flower with circular petals,

like this: Or like this:

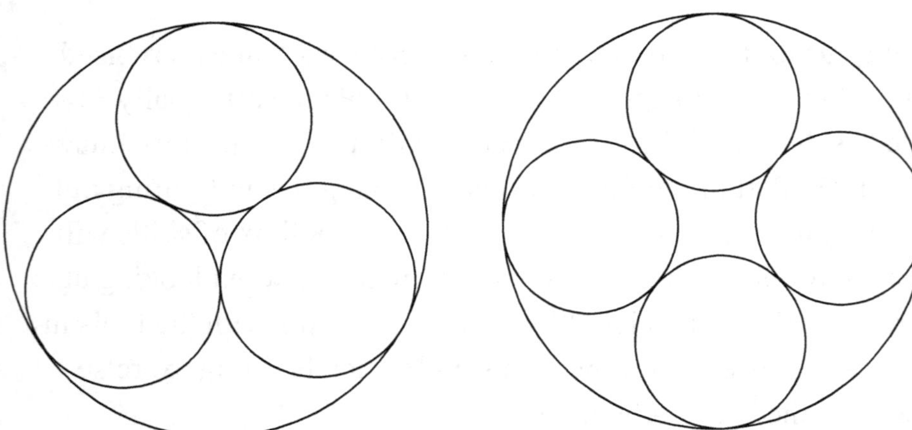

We can then ask some quantitative questions about it, like what's the radius of one of the inner circles, combined areas, circumferences, distances between centers of circles, angles between lines connecting centers and points of tangent, etc. Again, those will all be adventures that offer much more fun than the original plate painter "problem" in the textbook because they are asked through genuine curiosity with no initial regard to the skills required for the solution, as opposed to a "problem" built with the exact steps of the solution in mind.

But let's not stop here. Many agree that the spirit of mathematics, and the spirit of a *good* problem, one that gives us the satisfaction of having done some real, meaningful mathematics, usually lies in the underlying patterns that govern the specific cases. For example, does the combined area of the inner circles increase or decrease as we increase the number of petals? What about the combined circumference? Would the answers be different if we add in a central circle for the center of the flower? Would these patterns always hold true? Are there exceptions? And most importantly, *why*?

In *Thinking Mathematically*, Mason et al. (2010, pp. 8-9) call this act of looking at underlying patterns *the process of generalizing*. Together with its counterpart, *the process of specializing* (which deals with looking at specific cases to gain understanding), the two processes form the basis for mathematical thinking in both problem solving and problem making.

Of course, inspirations for cleanly abstract mathematical problems don't have to come strictly from abstract or synthetic situations. A friend and former colleague of mine, John Smith (J. Smith, personal communication, June 17, 2021), gave me this problem from the field of engineering after reading some rambling of mine on the process of problem making:

> What does a car "need to know" about itself and another car in order to parallel park behind that other car in such a manner that all the rules of the road are abided by? Does the length of either car matter or only the length of the car to be parked? Does the width of either car matter? When parallel parking behind a car of X meters in width, does the angle of entry increase or decrease (or remain unchanged) from the angle of entry needed when parking behind a car of (X-n) or (X+n) meters in width?
>
> Now that we have vehicles which will actually parallel-park behind another vehicle without the help of the driver, these are only a few problems that have to be solved for real. Yes, the cars are guided by laser beams and mechanical and electronic feedback systems within the vehicle, but a real live human being had to figure it all out in the first place, and probably starting with the question, "what does a car 'need to know' […] in order to parallel park?" I believe that any good math problem begins with simply *a problem.*

It stems from the natural process of encountering and attempting to solve real life problems all around us.

5.

Questions and Answers of Problem Making and Solving

By now you are probably feeling all kinds of uneasy. Your brain might be protesting with all kinds of questions. Let's answer some of them.

What if I'm just not a "problem making" person?

In case you missed it, I believe the vast majority of us are not at this point. We don't have the habit or the experience for it, unless we work for one of the annual contest organizers and are in charge of question creation. The whole purpose of this book is to change that, one non-problem-maker at a time, and it starts with you. As with learning anything in life, the key is in the persistent and deliberate practice, which begins with a willingness to try.

In her book *Mindset*, Dr. Carol Dweck (2008) talks about two differing beliefs about success. One is called the "fixed mindset", and it's the belief that our abilities are pre-determined by nature, and that no amount of practice can bring us above our "limit". In short, if I'm not a math person, I will struggle with math all my life regardless of what I do. The other is called the "growth mindset", which says that we can change our abilities, our "limits" by working on them. That is, if I'm not good at math *now*, I can become better at it by practicing, regardless of my innate mathematical "limit". The science is still unclear about which is closer to the truth, but believing one over the other makes a real and significant difference in the person's motivation. This results in people with a fixed mindset becoming unwilling to try new things or to work on things they struggle with, as well as assessment anxieties that stem from the fear of showing or finding their "limits".

For the sake of creating an elephant-less room for our future generations, please consider embracing the growth mindset on the issue of problem making and give it an earnest effort.

Are you saying we should stop doing formulae and drills altogether?

No, that's not what I'm saying at all. I know what you're referring to. Some believe that we should let students arrive at the formulae we intend to teach through their own exploration. That's not what I believe. There is a reason why some of these formulae took centuries for mathematicians to discover. Why do we need to have every student reinvent every wheel?

The idea of our new Core Competencies curriculum is to try to replace the boring *technical learning* with more elaborate *semantic learning*. It would help, but it wouldn't work completely. No matter how much we know about the structure and physics behind a hammer, its production procedures and material used, we still have to swing it onto a nail to become good at swinging hammers onto nails. Instead, I suggest we move through both *semantic* and *technical learning* quickly and get on with *contextual learning*.

When and how should we do the drills then?

Think of the topics in our current curriculum as tools to add to students' problem solving toolbox. When we want to give someone a hammer to add to their carpentry toolbox, we wouldn't say:

"Here's a hammer, now try hammering these 100 nails."

Instead, if they have never seen a hammer before, we would say:

"Here's a hammer for your toolbox. Here's how to use it. I hope it's helpful to you."

And they would thank us and add it to their toolbox. Then when they try to build a chair, they would take out the hammer and try to figure out how to use it. If they can't, they can then come back to us for help. (As teachers we would try to build in opportunities for this in our classroom routines.) Only then would it be the right time to say:

"Now try hammering these 100 nails."

This is because now they have a goal. They want to learn how to use the hammer so that they can build that chair. There is *contextual learning* waiting to happen after the completion of this *technical learning* session.

And even then we wouldn't force them to finish hammering all 100 nails. If they feel that they got the hang of it after 20, then they stop at 20 and go try the hammer on the chair they want to make. We don't have to give them a test on using the hammer either. This is because they can continue their practice on hammering in their chair project.

They can continue practicing what they haven't mastered in *technical learning* by using them in *contextual learning*.

What if they need a piece of mathematics we haven't covered in class to solve a problem they see?

Some pre-screening of the problems would help. I wouldn't give my square and triangle problem from the last chapter to elementary school classes for example. If I were to give out a problem, and it turns out to need some math beyond what we have learned so far, I would just teach the topic or ask students to research on the math they need. It is similar to when a student comes across a difficult word while reading a book. We wouldn't tell the student that it's a "grade 12 word" and is too hard for him. We would tell him to find a dictionary and look it up. This is much more meaningful of an introduction than "what comes next in the textbook". If it gets too complicated or too distracting from the problem, I would suggest that we leave it for now and finds another problem. It's no big deal.

What if they don't get to practice a particular technique we're supposed to teach in our curriculum in the problems they solve?

I believe the whole point of teaching multiple "techniques" is to give students a choice. There is no point forcing students to use multiple techniques when they are comfortable with one and using it in problem solving situations. If we are talking about an obscure skill from a small section of a chapter of the textbook, then we can either expand our problem set to include problems that require that skill or take a long hard look at why we still have that skill in our curriculum.

What if a student isn't the "problem solving type"? What if a student prefers drills over problems?

Throughout my time as a student in the K to 12 system, I have always believed that I'm not the "writing type" of person. I struggled with English as an ESL student from Grade 8, but even before having to do everything in an unfamiliar language, I struggled with Mandarin while growing up in China. It wasn't until recent years when I picked up creative writing as a hobby that I realized that the problem was never me being a specific "type". My struggle with writing was due to my lack of exposure to it all along.

If a "writing phobic" student like me can grow up to write a book, a "math phobic" elephant can float like a dandelion.

… I mean, a "math phobic" student can grow up to like problems.

Again, this goes back to promoting a growth mindset in the student. A person with a fixed mindset is more inclined to stick with familiar activities in fear of finding out that she is not good at the new activity. Therefore it's our job as the teacher to create a safe and inviting environment for problem solving.

How do we create such an environment?

Let's talk about the two types of motivation. When a person engages in an activity because the activity itself is fun, enjoyable, or rewarding, we say he is intrinsically motivated, as opposed to extrinsically motivated, which refers to when the person engages in an activity because of external factors, like marks, peer approval, or guilt. (Deci et al., 1985) Out of these two types, intrinsic motivation is associated with long-lasting and self-sustaining behaviors as well as greater mental health benefits. It is the preferred type of motivation for student learning.

What's interesting is that the presence of extrinsic motivators can decrease intrinsic motivation through a process called "overjustification". For example, rewarding students with stickers or marks for arriving at the correct final answer will contribute to *lowering* the students' enjoyment of the problem solving process itself, making students believe that they solve problems because they want the reward associated and not the enjoyment. We can start by *not* offering rewards for the correct final answer or the speed to get to it. In fact, take the focus off the final answer as much as we can. This reduces both the interference from extrinsic motivation and the possibility for "failure" that impedes students' willingness to try.

On the other hand, factors that help to increase intrinsic motivation are mastery (the feeling of getting better at something over time) and autonomy (being able to choose what to work on and how).

For mastery, we can encourage students to keep track of their thought process of solving each problem. This can build a sense of progress by showing students how much they have "wrestled" with the problem even when they haven't been successful at getting to the final answer. Mason et al. (2010) have some good suggestions on this in *Thinking Mathematically*. It doesn't have to be a formal mathematical solution, but writing down their thoughts will also help students learn how to communicate formally in mathematics down the road.

Then we want to give students freedom in both choosing the problems to work on and how to work on them. This gives them a sense of autonomy, which makes them feel in control of their progress and take ownership of their accomplishments.

Should we do problems as a whole class or individually?

In the spirit of giving students the freedom to choose problems they like and the way they want to solve them, I would suggest all problems start out as individual tasks. However, in problem solving, many people do find it beneficial to discuss progress and guesses with others. Even just the act of presenting our own thoughts to a peer forces us to think more clearly about our own solution, which might lead to breakthroughs. The two-stage setup for problem solving assignments, where students first work individually and then discuss in a small group, is worth exploring.

Wouldn't students be more likely to succeed at solving problems, and therefore like problem solving more, if we give out problems in the chapters used to solve them?

One of my favourite problems of all times was a problem I saw on the American Invitational Mathematics Examination in my Grade 11 year. (Mathematical Association of America, 2003) I don't remember the exact wording, but the problem was something like this:

> There are n points placed evenly around a circle. Starting with one of the points, we draw straight lines connecting them counterclockwise, skipping m points each step. When we return to the starting point, if we have connected all n points, then we have drawn a *star*. If some of the points are missed, it is not a *star*. *Stars* that can be rotated to look exactly the same

are considered the same *star*. How many unique *stars* can be drawn for n=100?

(Please stop reading here if you wish to try out this problem yourself. The next paragraph contains spoilers to the solution.)

It took me a while to figure out that this problem was not about drawing or circles or points at all. It was about relatively prime numbers. I was so excited when I figured it out that I still remember this problem almost two decades later. Now imagine if this problem was given to me after a chapter of relatively prime numbers in math class. It would've totally ruined my sense of accomplishment in identifying the right "tool" to use.

Please don't do this to your students. Don't present problems next to the tools you expect them to use. This is not *helping* them solve the problem. It's not *setting them up for success*. What you're doing is *ruining* the problem for them. It's called a *spoiler*.

What about assessments?

The good news is that we can still use the same tests we have on basic skills if that's what we are aiming to measure. We just need to think of them more as snap shots of the student's skills instead of an end point of learning. These test results can then be used as the basis for future problem recommendations or supplementary drills.

One of my proudest accomplishments while working as the Math Department Head of Wuhan Maple Leaf International School was the implementation of a system called Standard Based Grading in all our Calculus 12 classes. It's a system I would recommend for any *technical learning* based classroom, but I find the concept especially promising if we were to introduce *contextual* and *creative learning* into class time.

I first learned about Standard Based Grading from the blog of a teacher named Shawn Cornally, who has since taken down his online blog. We can still find information on it from blogs of teachers who were influenced by Shawn Cornally's ideas. An example is the blog of Lauren Stewart (2014) (available at https://modelsofar.wordpress.com/standards-based-grading/).

Standard Based Grading, or SBG, is a system of assigning and reporting grades on a student's learning progress that aims to be an accurate reflection of how much the student knows of the course material. How it differs from our traditional grade book layouts, which include items like assignments, quizzes, tests, exams, homework, participation, and bonus marks, is that we divide the course into measurable "standards" and grade students directly on the standards. Each recorded grade is a direct answer to the question "Does the student know this?" There are only three possible answers: Yes (2/2), no (0/2), or partially (1/2). What this means is that instead of recording "8/10 on Assignment 3" and "3/5 on Quiz 5" into the grade book, we would record "1/2 on Simplifying Fractions" and "2/2 on Two Digit Additions". The idea is that when we add these grades that link directly to the standards, the total is a direct reflection of how much the student knows. Not only that, the records plainly tell us which standards the students still need to work on.

Clarity is still not the biggest benefit of this system. Since we can easily see and change grades on a specific standard, it allows students to demonstrate learning at a later date on standards they haven't previously mastered. A student can initiate reassessment on a standard by asking for a quiz, for example, and if we determine that they really have learned, we can go into our grade book and change the record for that standard. The same would apply if we find that students have forgotten something they previously mastered. The grade book then becomes a dynamic snapshot of the students' learning progress. This takes away the inherent punishment for slow learners in the traditional marking system and encourages students to be responsible for their own learning by working on what they know they need to work on. Not only that, a large part of this process can be done by the students in their own time, so it can free up class time for *contextual* and *creative learning*.

If we do move on to assessing actual mathematical thinking in the future, if and when there is a curriculum update in that direction, I would imagine the assessment looking more like a rubric, similar to the ones English teachers use for assessing essays.

6.

The Idea of a Math Fair and a Community of Problem Makers

I came across the idea of a math fair during my teacher training at UBC. The professor of the Math Cohort, Dr. Susan Gerofsky, took us on a field trip to attend such an event hosted by a teacher named Chris Stroud and his Grade 7 class. The idea of it is absolutely charming. It's an event where students present their favourite problems in booths, helping visitors solve their problems by providing them with manipulatives and hints along the way. Students work in pairs so that they can take turns manning their own booth and visiting other students' booths and trying their problems. No solutions are provided at all to avoid spoiling the problems for visitors. The educators who designed the event called it the SNAP Math Fair, which stands for Student-centered, Noncompetitive, All-inclusive, Problem-based Math Fair. (More details can be found at www.mathfair.com.)

I was so in love with the idea that when I started my job at Wuhan Maple Leaf International School, I proposed that we run a math fair for all Math 10 and 11 students every semester. We put it into our math courses as an extra unit that didn't count towards their report card grade because these students, who spent the first nine years of their education in the highly competitive Chinese public school system, would breeze through our Grade 10 and 11 BC math curriculum, and we would run out of things to do toward the end of each semester.

The unit is about ten days long, with each day accomplishing a goal towards the actual math fair at the end of the unit.

On the first day they each select a partner and a problem from the Galileo Education Network's collection of math fair problems (available at https://galileo.org/math-fairs/math-fair-problems/).

On the second day they translated the problems into Mandarin because most of them were ESL students of two or three years, and sometimes they had trouble understanding long story-typed problems.

On the third day they solved their problems to the best of their abilities.

On the fourth day they translated their problems back into English using their own words, often giving the background stories their own spin. This would be the version to go on their problem posters.

On the fifth day they each wrote a slip of paper introducing themselves, and these slips would be included on their problem poster.

On the sixth day they made the manipulatives and the three hints that would help their visitors solve their problems.

On the seventh day they made their problem posters on sheets of A2 sized paper.

On the eighth day while they finished their posters, we had an interview with each pair of students to go over what would happen at the math fair and to make sure they can present their problems and hints verbally.

On the ninth day was an in-class practice run.

On the last day we put three or four classes of Math 10 and 11 students in a large dance room with their problems and invited the rest of the school to come and try them.

Everyone loved these math fairs. The math teachers loved them, their students loved them, and the visitors loved them. Every semester I would ask myself why I find all these extra tasks for myself and my colleagues to do leading up to the final day, but once the actual math fair got rolling, I would feel that it was worth all the trouble after all. There was only one issue. By the second year, when we had done them for the fourth time, people were getting tired of seeing the same problems. There were only about 80 of these problems, and people quickly learned them all after a few rounds.

The math fair as it is now is not sustainable as a regularly recurrent event of any substantial scale because of the shortage of good problems we have. In order for it to become sustainable, either we need to have a steady stream of good *new* problems from the adults to add to our pool, or preferably, we can have students *make* the problems they present in a math fair. This way, not only do they become experts at *solving* their problems, they *own* the problems, quite literally, and they would be presenting *their* problems, the products of their own curiosity and creativity. Not to mention we would always be seeing brand new problems at every math fair booth!

The question then becomes what *good math fair problems* look like and how to make them. The basic traits of a good math fair problem are very similar to those of a good problem in general, like the ones we talked about in Chapter 4.

- They are hard enough to require a bit of wrestling from their intended audience.
- They are not directly linked to specific skills.
- They look at underlying mathematical patterns instead of only specific cases.

There are also a few additional traits that make problems especially well suited for the math fair setting, making them easier to engage visitors who are casually walking by the booth.

- They come with parts that vary in difficulty and range from specific cases to more general investigations of underlying patterns.
- They are more compatible with physical manipulatives as a method of trial and error instead of pen-and-paper calculations.
- They can be put into a story setting to avoid complicated mathematical jargon.

There's something else we need to build before we all hole up with our math textbooks to cook up new and interesting math problems. I learned from my time spent in the creative writing community that creativity doesn't thrive in a vacuum. Not only does it feel lonely to create something and have no one to share it with, it's also too easy to run out of inspirations when we don't have the stimuli of other people's work. *Creative learning* needs to be built on a backdrop of *contextual learning*. This means we need a community of math problem makers.

Considering the current lack of popularity of problem making as a hobby, our best bet would probably be to build an online community. This way people who live apart but share this interest can come together and talk about their creations. Ideally this would be a database where problem makers can post their problems, and each problem has a comment section where people can have a discussion about it without posting the full solution. The database would need to be searchable by topic, intended grade level, difficulty, problem maker's account, and date posted for easy access to the problems.

I have in fact built an online space with some of these basic functionalities with the help of a more technologically inclined friend of mine. The website is still in its early phase, but you can now visit https://growadandelion.ca/category/math/ to browse original math problems (and other creative works on the site) or to share your own creations.

There are still things we can do until such online spaces become widespread. Start a side hobby of problem making today. Make problems. Try to solve your problems. (It's fine if you can't.) Share your problems with those around you: your family, friends, colleagues, students, neighbour's great aunt. Invite them to solve your problems. Tell them about the hobby of problem making. Invite them to become problem makers themselves. Start a problem makers' club at your school. Start an online community for problem makers. Join an online community for problem makers. Share your problems in your club or online. Solve other people's problems that are posted in your club or online and give them feedback. Publish your best problems in books. Host math fairs and invite people to share their problems. Join math fairs.

We need to all work together to build up the reservoir of authentic math problems our society has been missing, and then we can talk about teaching our students about problem solving.

I would like to close with thoughts on the purpose of all our efforts on eradicating the math phobia elephant from the room of education. The question of "how is math useful", the question that pertains to the science and application side of mathematics, has been asked loudly and clearly and thought about repeatedly in our school system. What has been excluded is the question of "what is there to experience in math", the question that draws to the artistic and expressive side of mathematics. As an art form, mathematics demands that we think of the question of "why we do math" not in terms of uses and benefits, but in terms of the human experiences it offers.

There is something unique in mathematics as a field of study outside of its scientific uses. It's a whole category of thought experiments that we don't get to run in other fields. Why do we do math? We do it for the same reason why we write stories or create art works or make music. These art forms are explorations that we human beings naturally do in our minds. It's not because they are beneficial that we do them. It's because they are a natural part of the human experience. It is true that not everyone needs to be a writer or an artist to have a fulfilling life, but everyone should at least be given the exposure and opportunity to experience them at some point in their life because they are such a fundamental form of human expression.

The art form of mathematics is the same. Every human being should experience it in its true glory at some point in their life. It is our responsibility as educators to provide such exposures and opportunities in our K-12 school system, a shared experience of all walks of life in our society. It's worth our every effort, and we need to get it right.

~ * ~

Reference List:

Books:

Bloom, B. S., Engelhart, M. D., Furst, E. J., Hill, W. H., Krathwohl, D. R. (1956). *Taxonomy of educational objectives: The classification of educational goals. Vol. Handbook I: Cognitive domain.* New York: David McKay Company.

Deci, E. L., Ryan, R. M. (1985). *Intrinsic motivation and self-determination in human behaviour.* New York: Plenum.

Dweck, C. S. (2008). *Mindset*. Ballantine Books.

Lockhart, P. (2009). *A Mathematician's Lament.* Bellevue Literary Press.

Mason, J., Burton, L., & Stacey, K.. (2010) Thinking Mathematically. (2nd ed.). Pearson.

Piaget, J. (1936). *Origins of intelligence in the child.* London: Routledge & Kegan Paul.

Online Sources:

BC's Curriculum. (n.d.) *Mathematics*. Province of British Columbia. https://curriculum.gov.bc.ca/curriculum/mathematics

Galileo Educational Network. (n.d.) *Math Fair Problems*. Galileo Education Network. https://galileo.org/math-fairs/math-fair-problems/

Grow a Dandelion. (n.d.) https://growadandelion.ca/category/math/

R.L.G. (2013, May 29). *Lexical facts.* The Economist. https://www.economist.com/johnson/2013/05/29/lexical-facts

SNAP Mathematics Foundation. *www.mathfair.com*

Stewart, L. (2015). *Standard Based Grading.* The Model So Far... https://modelsofar.wordpress.com/standards-based-grading/

Others:

Mathematical Association of America. (2003, March 25th). *2003 American Invitational Mathematics Examination I.* Mathematics Association of America

www.ingramcontent.com/pod-product-compliance
Lightning Source LLC
LaVergne TN
LVHW052102160826
845678LV00015B/3316

* 9 7 8 1 9 8 9 8 0 2 3 6 6 *